KU-778-567

PAT: Portable Appliance Testing
In-Service Inspection and Testing
of Electrical Equipment

Fourth Edition

Brian Scaddan

Routledge
Taylor & Francis Group
LONDON AND NEW YORK

Fourth edition published 2015
by Routledge
2 Park Square, Milton Park, Abingdon, Oxon OX14 4RN

and by Routledge
711 Third Avenue, New York, NY 10017

Routledge is an imprint of the Taylor & Francis Group, an informa business

© 2015 Brian Scaddan

The right of Brian Scaddan to be identified as author of this work has been asserted
by him in accordance with sections 77 and 78 of the Copyright, Designs and Patents Act
1988.

First edition published 2000 by Newnes, an imprint of Elsevier
Third edition published 2011 by Newnes, an imprint of Elsevier

British Library Cataloguing in Publication Data
A catalogue record for this book is available from the British Library

Library of Congress Cataloging-in-Publication Data
Scaddan, Brian.
PAT – portable appliance testing : in-service inspection and testing of electrical
equipment / Brian Scaddan. – 4th edition.
pages cm
Includes index.
1. Electric apparatus and appliances – Testing. 2. Electric machinery – Inspection.
3. Electric fault location. I. Title.
TK401.S28 2016
621.3028′7–dc23
2015000376

ISBN: 978-1-138-84929-7 (pbk)
ISBN: 978-1-315-72570-3 (ebk)

Typeset in Kuenstler 480 and Trade Gothic by
Servis Filmsetting Ltd, Stockport, Cheshire
Printed by Bell & Bain Ltd, Glasgow

MIX
Paper from
responsible sources
FSC® C007785

PAT: Portable Appliance Testing

- Fully up-to-date with the relevant parts of the 17th Edition IET Wiring Regulations: Amendment 3 and the 2012 Code of Practice
- Provides all the required information on portable appliance testing in a user-friendly manner
- Expert advice from an engineering training consultant, supported with colour diagrams, examples and tables

he Electricity at Work Regulations 1989 requires any electrical system to ﬔ constructed, maintained and used in such a manner as to prevent dan- ﬔr. This means that inspection and testing of systems, including portable ﬐pliances, is needed in order to determine if maintenance is required.

his book explains in clear language what needs to be done and includes ﬔpert advice on legislation as well as actual testing. The book contains an ﬔendix providing the electrical fundamentals needed by non-specialists ﬐ also has sample questions (with answers) for the C&G 2377 exam. It ﬐so an ideal revision guide for the non-specialist, such as maintenance ﬔf and caretakers who carry out these tasks part-time, alongside their any other duties.

ﬕian Scaddan, I Eng, MIET, is a consultant for and an Honorary Member ﬐City & Guilds with over 40 years' experience in Further Education and ﬔining. He is Director of Brian Scaddan Associates Ltd, an approved ﬕ & Guilds training centre offering courses on all aspects of electrical ﬔtallation contracting including the C&G 2382-15, 2392-10, 2377-22, ﬔ4-01, 2395-01 and 2396-01. He is also a leading author of books for ﬕe courses.

Contents

Preface

The introduction of The Electricity at Work Regulations (EAWR) 1989 prompted, among many other things, a rush to inspect and test *portable appliances*. The Regulations do not require such inspecting and testing, nor do they specifically mention portable appliances. They do, however, require any electrical system to be constructed, maintained and used in such a manner as to prevent danger, and in consequence inspection and testing of systems (portable appliances are systems) is needed in order to determine if maintenance is required.

All electrical equipments connected to the fixed wiring of an installation will need attention, not just portable appliances. I have, however, left the title of this book as *PAT: Portable Appliance Testing* as such words are now indelibly imprinted on our minds, even though it should read 'Inspection and Testing of In-Service Electrical Equipment'. PAT means an appliance tester that is portable, <u>not</u> a tester just for portable appliances!!

The book is intended for those who need to be involved in this inspection and testing process, either as a business venture or as an 'in-house' procedure to conform with the EAWR. It is also a useful reference document for anyone embarking on a City & Guilds 2377 course. The short answer questions at the end of each chapter are intended to test the readers' knowledge based on the content of this book. The sample City & Guilds type examination questions in Appendix 3 will need reference to the 4th edition of the Code of Practice for In-Service Testing of Electrical Equipment.

Brian Scaddan

CHAPTER 1

Legislation

Important terms used in this chapter:

- Electrical system
- Duty holder

By the end of this chapter the reader should:

- be aware of the legislation relevant to equipment testing,
- understand the meaning of an 'electrical system',
- know who a 'duty-holder' is and his/her responsibilities,
- be aware of the consequences of contravening the requirements of the EAWR 1989.

There are four main sets of legislation that are applicable to the inspection and testing of in-service electrical equipment:

- The Health and Safety at Work etc. Act (HSWA) 1974
- The Management of Health and Safety at Work Regulations (MHSWR) 1999, amended 2003
- The Provision and Use of Work Equipment Regulations (PUWER) 1998, amended 2002
- The Electricity at Work Regulations (EAWR) 1989.

THE HEALTH AND SAFETY AT WORK ETC. ACT (HSWA) 1974

This applies to all persons – employers and employees – at work, and places a duty of care on all to ensure the safety of themselves and others.

THE MANAGEMENT OF HEALTH AND SAFETY AT WORK REGULATIONS (MHSWR)1999

In order that the HSWA can be effectively implemented in the workplace, every employer has to carry out a risk assessment to ensure that employees, and those not in his/her employ, are not subjected to danger.

PAT: Portable Appliance Testing: In-Service Inspection and Testing of Electrical Equipment. 978-1-138-84929-7. © Brian Scaddan. Published by Taylor & Francis. All rights reserved.

THE PROVISION AND USE OF WORK EQUIPMENT REGULATIONS (PUWER) 1998

Work equipment must be constructed in such a way that it is suitable for the purpose for which it is to be used. Once again, the employer is responsible for these arrangements.

THE ELECTRICITY AT WORK REGULATIONS (EAWR) 1989

Regulation 16 of EAWR 1989 should be mentioned. This Regulation is absolute; this means no matter what the time or cost involved, it must be done. This Regulation deals with the person being competent. The only way to prove to a court of law that you are a competent person is through evidence of regular training. Regular training? Every week or perhaps when new Regulations are brought out?

These regulations, in particular, are very relevant to the inspection and testing of in-service electrical equipment. There are two important definitions in the EAWR:

1. the electrical system
2. the duty holder.

Note

Although the IET Wiring Regulations BS 7671: 2008 Amendment 3 are non-statutory, it should be established that the fixed wiring of an installation is in a suitably safe condition for the connection of electrical equipment.

Electrical system

This is anything that generates, stores, transmits or uses electrical energy, from a power station to a wrist-watch battery. The latter would not give a person an electric shock, but could explode if heated, giving rise to possible injury from burns.

Duty holder

This is anyone (employer, employee, self-employed person, etc.) who has 'control' of an electrical system. Control in this sense means designing, installing, working with or maintaining such systems. Duty holders have a legal responsibility to ensure their own safety and the safety of others whilst in control of an electrical system.

The EAWR do not specifically mention inspection and testing; they simply require electrical systems to be 'maintained' in a condition so as not to cause danger. However, we only know if a system needs to be maintained if it is inspected and tested, and thus the need for such inspection and testing of a system is implicit in the requirement for it to be maintained.

Anyone who inspects and tests an electrical system is, in law, a duty holder and must be competent to undertake such work.

PROSECUTIONS

Offences committed under The EAWR 1989 may be liable for: £20 000 fine for each offence in Magistrates' Court, unlimited fines/prison sentences in Crown Court.

Here are just a few examples of the many prosecutions under the EAWR 1989 that take place every year.

Case 1.1

A greengrocer was visited, probably for the second time, by the Health and Safety Executive inspectors, who found 11 faults with the electrical installation. They were:

1. a broken fuse to a fused connection unit;
2. a broken three-way lighting switch;
3. a broken double socket outlet;
4. a broken bayonet light fitting;
5. a missing ceiling rose cover;
6. the flexible cable feeding the beetroot boiler went under the casing and not through the proper hole in the side;
7. there was no earthing to a fluorescent fitting;
8. there was no earthing to a metal spotlight;

9. block connectors were used to connect some bulkhead lights;
10. block connectors were used to connect the fluorescent lights;
11. block connectors were used to connect a spotlight.

He was subsequently fined £4950, and although he was 'only a green-grocer', he was also a duty holder, and as such had a responsibility for the safety of the staff working in the shop.

Case 1.2

An electrician received serious burns to his face, arms and legs after he was engulfed in a ball of flames whilst testing an old motor control switch-board. He was reaching into the board to test contacts located only a few inches away from exposed, live, 400 V terminals when the accident happened. He was apparently using inappropriate test leads that were unfused and had too much exposed metal on the tips. He was also working near live terminals because no arrangement had been made for the board to be made dead.

His company was fined a total of £1933 because they did not prevent work on or near live equipment. They were duty holders. The electrician, however, also a duty holder, carried the main responsibility for the accident, but would not have been prosecuted, as he was the only one to be injured.

Case 1.3

A young foreman on a large construction site was electrocuted when he touched the metal handle of a site hut which had become live. An employee of the company carrying out the electrical contracting work on the site had laid inadequate wiring in the hut which had later been crushed by its weight, causing a fault. Consequently the residual current device (RCD) protecting the hut kept tripping out, as it should have. However, another of the electrical contractor's employees by-passed the RCD so that it would not trip. This caused the site hut to become live.

The construction company was fined £97 000 for failing to monitor site safety, the electrical contractors were fined £30 000 and the contractor's managing director was fined £5000 and disqualified from being a company director for 3 years.

Questions

1. In UK legislation, which legal document relates to inspection and testing of electrical equipment?
2. Who is responsible, in an organisation, for compliance with the 'Provision and Use of Work Equipment Regulations?
3. What would a 230 V hair dryer be defined as, according to the EAWR 1989?
4. What is the title, in law, given to a person carrying out inspection and testing of electrical equipment?
5. Under what circumstances could a contravention of the EAWR 1989 result in a prison sentence?

Answers

1. The EAWR 1989
2. An employer
3. An electrical system
4. A duty holder
5. When the trial is in Crown Court.

CHAPTER 2

Setting Up

Important terms used in this chapter:

- Duty holder
- Responsible person

By the end of this chapter the reader should:

- understand the duties of a 'responsible person',
- be aware of the need to organize, monitor and record the results of an inspection and testing regime,
- be able to make decisions regarding the frequency and type of inspection and testing to be conducted,
- know what documentation need to be completed.

There are two ways for an organization to ensure that in-service electrical equipment is regularly maintained:

- employ a specialist company to provide the inspection and testing service; or
- arrange for 'in-house' staff to carry out the work through relevant training to ensure competence and hence compliance with Regulation 16 of the EAWR.

In either case, the first step is for the organization to appoint a 'responsible person' who will, therefore, be a duty holder and to whom staff and/or outside contractors should report the results of any inspection and test, including defects, etc. Such a person could be the manager of the premises or a member of staff: they will need to be trained and competent, both in the management of the appliance testing process and in the knowledge of relevant legislation as discussed in Chapter 1.

The second step is for the 'responsible person' to carry out an inventory of all equipments that will need testing and/or inspecting, and make decisions as to the frequency of such work. Some advice may be needed here from an experienced contractor in order to achieve the most effective time schedule and to make decisions on which equipment should be involved.

PAT: Portable Appliance Testing: In-Service Inspection and Testing of Electrical Equipment.
978-1-138-84929-7. © Brian Scaddan. Published by Taylor & Francis. All rights reserved.

Table 2.1 gives some examples of recommended periods between each inspection and test.

The 'responsible person' should have in place a procedure for users of electrical equipment to report and log any defects found.

Table 2.1	Sample of Suggested Frequencies of Inspection and Testing				
Equipment	Class	Inspection and Tests	Offices and Shops	Hotels	Schools
Hand-held	Class I and II	User checks	Before use	Before use	Before use
	Class I	Formal visual inspection	Every year	Every year	Every 6 months
		Combined inspection and test	Every 2 years	Every 2 years	Every year
	Class II	Formal visual inspection	Every year	Every year	Every year
		Combined inspection and test	None	None	Every 4 years
Portable	Class I and II	User checks	Weekly	Weekly	Weekly
	Class I	Formal visual inspection	Every year	Every year	Every 6 months
		Combined inspection and test	Every 2 years	Every 2 years	Every year
	Class II	Formal visual inspection	Every 2 years	Every 2 years	Every year
		Combined inspection and test	None	None	Every 4 years
Moveable	Class I and II	User checks	Weekly	Weekly	Weekly
	Class I	Formal visual inspection	Every year	Every year	Every 6 months
		Combined inspection and test	Every 2 years	Every 2 years	Every year
	Class II	Formal visual inspection	Every 2 years	Every 2 years	Every year
		Combined inspection and test	None	None	Every 4 years

Table 2.1 Sample of Suggested Frequencies of Inspection and Testing—Cont'd

Equipment	Class	Inspection and Tests	Offices and Shops	Hotels	Schools
Stationary	Class I and II	User checks	None	None	Weekly
	Class I	Formal visual inspection	Every 2 years	Every 2 years	None
		Combined inspection and test	Every 5 years	Every 5 years	Every year
	Class II	Formal visual inspection	Every 2 years	Every 2 years	Every year
		Combined inspection and test	None	None	Every 4 years
IT	Class I and II	User checks	None	None	Weekly
	Class I	Formal visual inspection	Every 2 years	Every 2 years	None
		Combined inspection and test	Every 5 years	Every 5 years	Every year
	Class II	Formal visual inspection	Every 2 years	Every years	Every year
		Combined inspection and test	None	None	Every 4 years

Whether the inspection and test is to be carried out by competent staff or by outside contractors, it is advisable that various forms be produced.

EQUIPMENT REGISTER

This details equipment that may need to be inspected and tested (Figure 2.1).

COMBINED INSPECTION AND TESTING FORM

This details the results of formal visual inspection or combined inspection and testing (Figure 2.2).

Equipment Register						
COMPANY: *Jones Footware Ltd., Blacktown.*						
					Frequency of Insp. & Test	
Register No.	Equipment	Equip. No.*	Class I, II or III	Normal Location	Formal visual Insp.	Combined Insp. & Test
1	Kettle	12	I	Kitchen	6 mths.	12 mths.
2						
3						
4						
5						
6						
7						

* This could be the serial No. or a number allocated by the company or the contractor and durably marked on the equipment

FIGURE 2.1   Equipment register.

FAULTY EQUIPMENT AND REPAIR REGISTER

This details faulty equipment taken out of service and sent for repair (Figure 2.3).

Previous records must be kept and made available to any person conducting routine inspection and testing of in-service electrical equipment.

Inspection and Testing Record

COMPANY: *R.F. Bloggins & Son Ltd., Whiteford*

Equipment	Equip. No.	Class I, II or III	Normal location
Floor polisher	*8*	*1*	*Store room*

Make:	*Lynatron*	Purchase date: *1.2.2007*
Model:	*KPX2*	
Serial No:	*13579*	

Voltage:	*230 V*
Power:	*700 W*
Current:	*N/A A*
Fuse:	*5 A*

Frequency of inspection and testing	
Formal visual	*Weekly*
Combined insp. & test	*12 mths*

	Inspection						Testing						
	Correct environment for use	Permission to disconnect*	Socket	Plug	Flex	Body	Earth continuity		Insulation resistance	App. Voltage E. Leakage		Functional	OK to use	Signature
Date							Ohms	OK	M-ohms	mA			
1.2.2008	*Yes*	*N/A*	*OK*	*OK*	*OK*	*OK*	*0.07*	*Yes*	*200 +*	*N/A*	*OK*	*YES*	*A. Munn*
8.2.2008	*Yes*	*N/A*	*OK*	*OK*	*OK*	*OK*						*YES*	*A. Munn*
15.2.2008	*Yes*	*N/A*	*OK*	*OK*	*OK*	*OK*						*YES*	*A. Munn*
22.2.2008	*Yes*	*N/A*	*OK*	*OK*	*OK*	*OK*						*YES*	*A. Munn*

* Applies to business and IT equipment which may need downloading first.

FIGURE 2.2 Inspection and testing record.

Faulty Equipment & Repair Register

COMPANY: Mr. Baldys Hairdressing Emporium, Thintown

Date removed from service	Equipment	Equip. No.	Equipment register No.	Normal location	Fault	Date sent for repair	Repairer	Date returned	Suitable for use OK	Suitable for use Signature	Comments
13.3.2008	Hair dryer	9	4	Main salon	Frayed flex	20.3.2008	N.O. Good	28.3.2008	Yes	*N.O. Good*	
15.3.2008	Curling tongs	11	18	Room 2	Cracked handle	20.3.2008	T.O. Bad	1.4.2008	No	*T.O. Bad*	Not repairable

FIGURE 2.3 Faulty equipment and repair register.

Questions

1. What are the main duties of a duty holder?
2. How often should a combined inspection and test be carried out on Class I stationary equipment in a hotel?
3. In an organization, what are the PAT documents that need to be available for completion?
4. What should happen to the records of equipment inspection and testing?

Answers

1. Carry out equipment inventories. Liaise with staff and those conducting the inspection and testing. Receive reports and take action with regards to faulty equipment.
2. Every 5 years.
3. Equipment register; combined inspection and test report form; faulty equipment register.
4. Must be kept and made available to persons carrying out future inspections and tests.

CHAPTER 3

Equipment to be Inspected and Tested

Important terms used in this chapter:

- Basic protection
- Fault protection

By the end of this chapter the reader should:

- be able to distinguish between classes of equipment,
- recognize symbols found on Class II and Class III equipment,
- recognize various equipment types.

As mentioned in the Preface to this book, it is not just portable appliances that have to be inspected and tested, but all in-service electrical equipment. This includes items connected to the supply by BS 1363 13 A plugs, BS EN 60309-2 industrial plugs or hard wired to the fixed installation via fused connection units or single- or three-phase isolators.

It is perhaps wise at this stage to comment on the two methods of protecting against an electric shock, and the different classes of equipment (Class 0, Class 01, Class I, Class II and Class III).

BASIC PROTECTION

This prevents touching intentionally live parts. Protection is generally achieved by applying basic insulation to such parts and/or enclosing them to prevent contact.

FAULT PROTECTION

This provides protection where exposed metalwork of electrical equipment has become live due to a fault (e.g. breakdown of basic insulation).

PAT: Portable Appliance Testing: In-Service Inspection and Testing of Electrical Equipment. 978-1-138-84929-7. © Brian Scaddan. Published by Taylor & Francis. All rights reserved.

Protection is generally by adequate earthing and automatic disconnection of supply or the use of double or reinforced insulation (Class II).

CLASS 0 EQUIPMENT OR APPLIANCES

Almost everyone can remember those old-fashioned, ornate brass table lamps, wired with either flat PVC-insulated twin flex or twisted cotton-covered rubber-insulated twin flex. In other words, equipment with a non-earthed metal case. Protection against electric shock is only provided by insulating live parts with basic insulation. Breakdown of this insulation could result in the metal enclosure becoming live and with no means of disconnecting the fault. The statutory Electrical Equipment Safety Regulations introduced in 1975 effectively ban the sale of Class 0 equipment.

CLASS 01 EQUIPMENT OR APPLIANCES

This is the same as Class 0. However, the metal casing has an earthing terminal but the supply cable is twin and the plug has no earth pin.

Class 0 equipment is not acceptable in the UK and class 01 may only be used in special circumstances and in a strictly controlled environment. Generally, these classes should not be used unless connections to earth are provided on the item and an earth return path via a supply cable has a circuit protective conductor (cpc) incorporated: this would convert the equipment to Class I.

CLASS I EQUIPMENT OR APPLIANCES

These items have live parts protected by basic insulation and a metal enclosure or accessible metal parts that could become live in the event of failure of the basic insulation. Protection against shock is by basic insulation and earthing via casing, the cpc in the supply cable and the fixed wiring of the installation.

Typical Class I items include toasters, kettles, washing machines, lathes and pillar drills (see Figures 3.1 and 3.2).

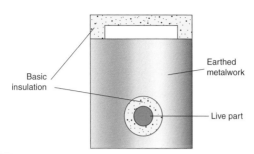

FIGURE 3.1 Class I equipment.

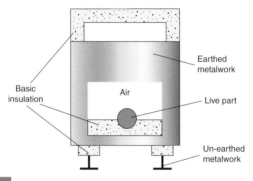

FIGURE 3.2 Class I equipment.

CLASS II EQUIPMENT OR APPLIANCES

Commonly known as double-insulated equipment, the items have live parts encapsulated in basic and supplementary insulation (double), or one layer of reinforced insulation equivalent to double insulation (Figures 3.3 and 3.4).

Even if the item has a metal casing (for mechanical protection) it does not require earthing as the strength of the insulation will prevent such metalwork becoming live under fault conditions. The cable supplying such equipment will normally be two core with no cpc (Figure 3.5).

Examples of Class II equipment would include most modern garden tools such as hedge trimmers and lawn mowers and also food mixers, drills,

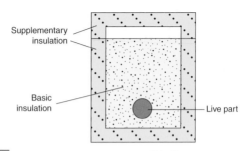

FIGURE 3.3   Class II equipment.

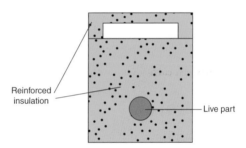

FIGURE 3.4   Class II equipment.

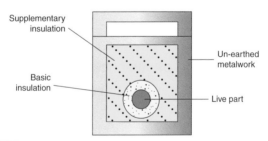

FIGURE 3.5   Class II equipment.

table lamps, etc. All such items should display the Class II equipment symbol:

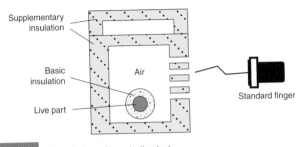

Supplementary insulation

Basic insulation

Air

Live part

Standard finger

FIGURE 3.6 Manufacturers' penetration test.

Equipment with grilles or openings (e.g. hair dryers) needs to pass the standard finger entry test carried out by the manufacturer (Figure 3.6).

CLASS III EQUIPMENT OR APPLIANCES

These are equipment/appliances that are supplied from a Separated Extra Low Voltage (SELV) source, which will not exceed 50 V and are usually required to be less than 24 or 12 V. Typical items would include telephone answer machines and other items of IT equipment. Such equipment should be marked with the symbol:

and be supplied from a safety isolating transformer to BS EN 60742 or BS EN 61558-2-6, which in itself should be marked with the symbol:

These transformers are common and are typical of the type used for charging mobile phones, etc. Note that there are no earths in an SELV system and hence the earth pin on the transformer is plastic.

EQUIPMENT TYPES

The Code of Practice for In-Service and Testing of Electrical Equipment defines various types of equipment/accessory that need to be inspected and tested and that are generally in normal use. Advice from the manufacturer should be sought before testing specialist equipment. The equipment types are as follows:

- portable equipment/appliances
- hand-held equipment/appliances
- moveable equipment/appliances
- stationary equipment/appliances
- fixed equipment/appliances
- built-in equipment/appliances
- information technology equipment
- extension leads.

Portable equipment/appliances

These are items which are capable of easy movement whilst energized and/or in operation. Examples of such appliances are:

- chip fryers
- toasters
- coffee percolators
- tin openers.

Hand-held equipment/appliances

These items are of a portable nature, which require control/use by direct hand contact. Examples include:

- drills
- hair dryers
- hedge trimmers
- soldering irons.

Moveable equipment/appliances

There is a thin dividing line between this and the previous two types, but in any case it still needs inspecting and testing. Generally such items are 18 kg or less and may have wheels or are easily moved. Examples would include:

- tumble dryers
- the old-fashioned twin-tub washing machine
- industrial/commercial kitchen equipment.

Stationary equipment/appliances

These appliances weigh in excess of 18 kg and are not intended to be moved, such as:

- ordinary cookers
- dishwashers
- washing machines.

Fixed equipment/appliances

These items are fixed or secured in place, typically:

- tubular heaters
- lathes and other industrial equipment
- towel rails.

Built-in equipment/appliances

These are appliances that are 'built-in' to a unit or recess, such as:

- an oven
- an inset electric fire.

Information technology equipment

In general terms, these are business equipment such as:

- PCs
- printers

- typewriters
- scanners.

Extension leads

These include the multi-way sockets so very often used where IT equipment is present, as there are seldom enough fixed socket outlets to supply all the various units. These leads should always be wired with three-core (line, neutral and earth) cable, and should not exceed:

- 12 m in length for a 1.25 mm^2 core size
- 15 m in length for a 1.5 mm^2 core size
- 25 m in length for a 2.5 mm^2 core size.

The 2.5 mm^2 lead should be supplied via a BS EN 60309-2 plug, and if any of the lengths are exceeded, the leads should be protected by a BS 7071 30 mA RCD.

Questions

1. What is provided by (a) Basic Protection and (b) Fault Protection?
2. What Class of equipment is banned by the 'Electrical Equipment Safety Regulations' 1975?
3. What Class of equipment with a metal case does not need the case to be earthed?
4. What is the symbol shown on Class II equipment?
5. What test is carried out by the manufacturer on equipment that has openings or grilles?
6. What are the conditions for an item of equipment to be classified as 'stationary'?
7. What protection should be provided for a 20 m long, 1.5 mm^2 extension lead?

Answers

1. (a) Protection against contact with live parts.
   (b) Protection against contact with parts made live because of a fault.
2. Class 0.
3. Class II.

4.

5. Standard finger penetration test.
6. Over 18 kg and not meant to be moved.
7. 30 mA RCD protection.

CHAPTER 4

Inspection

Important terms used in this chapter:

- User checks
- Formal visual inspection

By the end of this chapter the reader should:

- know when user checks and formal visual inspections need to be carried out,
- know what needs checking,
- know what records should be kept,
- know what action should be taken when faulty equipment is identified,
- be able to select appropriate cable sizes relevant to equipment power ratings.

Inspection is vital and must precede testing. It may reveal serious defects which may not be detected by testing only.

Two types of inspection are required: user checks and formal visual inspection.

USER CHECKS

All employees are required by the Electricity at Work Regulations to work safely with electrical appliances/equipment and hence all should receive some basic training/instruction in the checking of equipment before use. (This training needs to be only of a short duration.) Generally, this is all common sense: nevertheless, a set routine of pre-use checks should be established. Such a routine could be as follows:

- Check the condition of the appliance/equipment (look for cracks or damage).
- Examine the cable supplying the item, looking for cuts, abrasions, cracks, etc.
- Check that the cable sheath is secure in the plug and the appliance.

PAT: Portable Appliance Testing: In-Service Inspection and Testing of Electrical Equipment.
978-1-138-84929-7. © Brian Scaddan. Published by Taylor & Francis. All rights reserved.

- Look for signs of overheating.
- Check that it has a valid label indicating that it has been formally inspected and tested and the date of the next inspection and/or test.
- Decide if the item is suitable for the environment in which it is to be used, for example 230 V appliances should not be used on a construction site, unless protected by a 30 mA RCD.
- If all these checks prove satisfactory, check that the appliance is working correctly.

If the user feels that the equipment is not satisfactory, it must be switched off, removed from the supply, labelled 'Not to be used' or words to that effect, and reported to a responsible person. That person will then take the necessary action to record the details of the faulty item and arrange for remedial work or have it disposed of.

NOTE: No record of user checks is required if the equipment is considered satisfactory.

FORMAL VISUAL INSPECTION

This must be carried out by a person competent to do so, and recorded on an appropriate form. This inspection is similar to, but more detailed than, user checks and must be conducted with the accessory/equipment disconnected from the supply.

General

- Check cable runs to ensure that cables will not be damaged by staff or heavy equipment.
- Make sure that plugs, sockets, flex outlets, isolators, etc., are always accessible to enable disconnection/isolation of the supply, either for functional, maintenance or emergency purposes. For example, in many office environments, socket outlets are very often obscured by filing cabinets, etc.
- Check that items that require clear ventilation, such as convector heaters, VDUs, etc., are not covered in paper, files, etc., and that foreign bodies or moisture cannot accidentally enter such equipment.

- Ensure that cables exiting from plugs or equipment are not tightly bent.
- Check that multi-way adaptors/extension leads are not excessively used.
- Check that equipment is suitable for both the purpose to which it is being put and the environment in which it is being used.
- Ensure that accessories/equipment are disconnected from the supply during the inspection process, either by removing the plug or by switching off at a connection unit or isolator.
- Take great care before isolating or switching off business equipment. Ensure that a responsible person agrees that this may be done, otherwise this may result in a serious loss of information, working processes, etc.

The accessories/equipment

- Check the cable for damage. Is it too long or too short?
- Is the supply cable/cord to the appliance the right size?
- Is the plug damaged? Look for signs of overheating, etc.
- Is the fuse in a BS 1363 13 A plug the correct size? Are the contacts for the fuse secure? This requires dismantling of the plug. The fuse should be approved, and ideally have an ASTA mark on it. Some fuses made in China and marked PMS are dangerous and should be replaced. Fuse and cable sizes (in accordance with BS 1363) in relation to appliance rating are, in general, shown in Table 4.1.
- If a plug is damaged and is to be replaced, ensure that the replacement has sleeved live pins. The Plugs and Sockets etc. Regulations 1994 makes it illegal to sell plugs without such sleeved pins. However, this requirement is not retrospective, in that it does not apply to plugs with unsleeved pins already in use.

Table 4.1 Standard Flexible Cable Sizes

Appliance Rating	Cable Size
700–1300 W	0.75 mm^2
1300–2300 W	1 mm^2
2300–3000 W	1.25 mm^2

Questions

1. What action should be taken, regarding the supply to equipment, before inspection and testing is carried out?
2. What action should be taken if the user of the equipment considers that it is unsafe?
3. During a formal visual inspection, what action is necessary if a 13 A plug has unsleeved pins?
4. What flexible cable conductor size is appropriate for a 2.0 kW kettle?
5. What procedure should be taken before disconnecting business/IT equipment for the purposes of inspection and testing?

Answers

1. Items should be disconnected from the supply.
2. Take out of service and report to the responsible person.
3. No action is needed unless it is damaged; then replace with the one with sleeved pins.
4. 1.0 mm^2.
5. Liaise with and seek permission of the responsible person.

CHAPTER 5

Combined Inspection and Testing

Important terms used in this chapter:

- Earth continuity
- Insulation resistance
- Functional tests
- Flash or dielectric strength tests

By the end of this chapter the reader should:

- understand why inspection should precede testing,
- be aware that equipment should be isolated from the supply before testing,
- know the types of test equipment that can be used,
- know the tests to be carried out and how they are conducted,
- understand the reasons for 'earth continuity' tests,
- be able to interpret the test results and make relevant adjustments, if necessary,
- understand the reasons for 'insulation resistance' tests,
- be aware of the minimum values of insulation resistance for various classes of equipment,
- be aware of the danger of conducting 'flash/dielectric strength' tests.

Combined inspection and testing comprises preliminary inspection as per Chapter 4 together with instrument tests to verify earth continuity, insulation resistance, functional checks and, in the case of cord sets and extension leads, polarity as well. In some low-risk environments such as offices, shops, hotels, etc., Class II equipment does not require the routine instrument tests.

TESTING

This has to be carried out with the appliance/equipment isolated from the supply. Such isolation is, of course, easy when the item is supplied via a plug and socket, but presents some difficulties if it is permanently

PAT: Portable Appliance Testing: In-Service Inspection and Testing of Electrical Equipment.
978-1-138-84929-7. © Brian Scaddan. Published by Taylor & Francis. All rights reserved.

wired to, say, a flex outlet, a connection unit, or an isolator, etc. In these cases, the tester must be competent to undertake a disconnection of the appliance; if not, then a qualified/competent electrical operative should carry out the work.

Additionally, the permission of a responsible person may be needed before isolating/disconnecting business equipment.

PRELIMINARY INSPECTION

This must always be done before testing as it could reveal faults that testing may not show, such as unsecured cables in appliance housings, damaged cable sheathing, etc. The inspection procedure is as detailed in Chapter 4.

Testing

This may be carried out using a portable appliance tester, of which there are many varieties, or separate instruments capable of measuring continuity and insulation resistance.

Portable appliance testers

These instruments allow appliances, fitted with a plug, to be easily tested. Some testers have the facility for testing appliances of various voltage ranges, single and three phase, although the majority only accept single-phase 230 or 110 V plugs (BS 1363 and BS EN 60309-2).

Generally, portable appliance testers are designed to allow operatives to 'plug in' an item of equipment, push a test button, view results and note a 'pass' or 'fail' indication. The operative can then interpret these results and, where possible, make adjustments which may enable a 'fail' indication to be changed to a 'pass' status.

Some portable appliance testers are of the GO, NO-GO type, where the indication is either a red (fail) or green (pass) light. As there are no test figures associated with this type of tester, no adjustment can be made. This could result in appliances being rejected when no fault is present. This situation will be dealt with a little later.

Continuity/insulation resistance testers

These are usually dual instrument testers, although separate instruments are in use. Multi-meters are rarely suitable for these tests.

For earth continuity, the instrument test current (a.c. or d.c.) should be between 20 and 200 mA with the source having an open-circuit voltage of between 100 mV and 24 V. For insulation resistance, the instrument should deliver a maintainable test voltage of 500 V d.c. across the load. *Note*: All test leads should conform to the recommendations of the HSE Guidance Note GS 38.

So, what are the details of the tests required?

Earth continuity

This test can only be applied to Class I equipment, and the purpose of the test is to ensure that the earth terminal of the item is connected to the casing effectively enough to result in the test between this terminal and the casing giving a value of not more than 0.1 Ω.

Clearly, it is not very practicable to have to access terminals inside an enclosure and hence it is reasonable to measure the earth continuity from outside, via the plug and supply lead. This also checks the integrity of the lead earth conductor, or cpc.

Testing in this way will, of course, add the resistance of the lead to the appliance earth resistance, which could result in an overall value in excess of the 0.1 Ω limit, and the tester may indicate a 'fail' status. This is where the interpretation of results is so important in that, provided the final value having subtracted the lead resistance from the instrument reading is no more than 0.1 Ω, the appliance can be passed as satisfactory.

The use of a GO, NO-GO instrument prohibits such an adjustment as there are no test values available. Table 5.1 gives the resistance in ohms per metre of copper conductors, at 20 °C for flexible cords from 0.5 to 4.0 mm^2.

Hence, the cpc of 5 m of 1.0 mm^2 flexible cable would have a resistance of:

$$5 \times 0.0195 = 0.0975 \ \Omega$$

Table 5.1   Conductor Resistances

Conductor Size (mm²)	Resistance (Ω/m)
0.5	0.039
0.75	0.026
1.0	0.0195
1.25	0.0156
1.5	0.013
2.5	0.008
4.0	0.005

It is unlikely that appliances in general use will have supply cords in excess of $1.25\,\text{mm}^2$ as the current rating for such a cable is 13 A, which is the maximum for a BS 1363 plug.

Example 5.1

The measured value of earth continuity for an industrial floor polisher, using a portable appliance tester, is $0.34\,\Omega$. The supply cable is 10 m long and has a conductor size of $0.75\,\text{mm}^2$. The test instrument also indicates a 'fail' condition. Can the result be overruled?

Resistance of cpc of lead $= 10 \times 0.026 = 0.26\,\Omega$

Test reading, less lead resistance $= 0.34 - 0.26 = 0.08\,\Omega$

This is less than the maximum of $0.1\,\Omega$, so, yes, the appliance is satisfactorily earthed, and the test reading can be overruled to 'pass'.

The only problem with this approach is that most portable appliance testers have electronic memory which can be downloaded to software on a PC, which would record $0.34\,\Omega$ and hence a 'fail' status. Unless the instrument or the software includes the facility to include lead resistance, the appliance still fails (something to be said for paper records?).

Having made the above comments, it must be said that only low-power appliances with very long cables having small size conductors cause any problems.

Conducting the earth continuity test

Portable appliance tester

Having conducted the preliminary inspection:

- Plug the appliance into the tester and select, if possible, a suitable current. This will be 1.5 times the fuse rating (if the correct fuse is in place) up to a maximum of 25 A.
- Connect the earth bond lead supplied with the tester to a suitable earthed point on the appliance. (Remember that just because there is metal, it does not mean that it is connected to earth.) A fixing screw securing the outer casing to a frame is often the best place, rather than the actual casing, which may be enamelled or painted and may contribute to a high-resistance reading. If a high reading is obtained, other points on the casing should be tried.
- Start the test and record the test results.
- Do not touch the appliance during the test.

Figure 5.1 illustrates such a test.

Continuity tester

The method is, in general, as for the portable appliance tester but in this case a short circuit current of between 2 mA and 200 mA is delivered by the instrument and it is known as the 'soft test':

- Zero the instrument.
- Connect one lead to the earth pin of the plug.

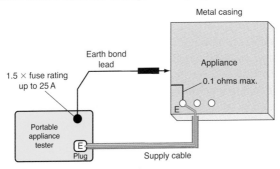

FIGURE 5.1 Continuity test with portable appliance tester.

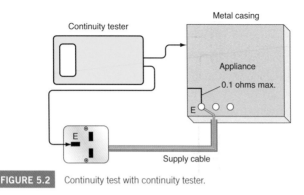

FIGURE 5.2   Continuity test with continuity tester.

- Connect the other lead to the appliance casing.
- Start the test and record the test results.
- Do not touch the appliance during the test.

Figure 5.2 illustrates such a test.

For the purpose of conducting an earth continuity test using a separate instrument, it would be useful to construct a simple means of 'plugging-in' and measuring, rather than trying to make contact with plug pins using clips or probes.

The resourceful tester will make up his/her own aids to testing. Such an aid in this case could be a polypropylene box housing a 13 A and a 110 V socket, with the earth terminals brought out to a metal earth stud suitable for the connection of a test lead (Figure 5.3).

Again, in the case of testing items of equipment that have to be disconnected from the supply, special test accessories are useful to aid the testing process. Such an accessory would be, for example, a plug, short lead and connector unit, to which a disconnected item could be connected. This is especially useful when using a portable appliance tester, whereas a continuity tester can be connected easily to the exposed protective conductor of the equipment.

Multi-way extension sockets and extension leads are to be treated as Class I equipment. However, there is some difficulty in gaining a connection to

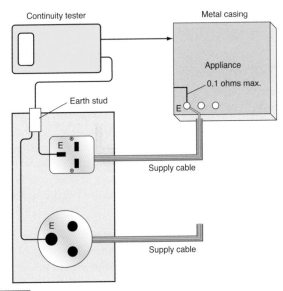

FIGURE 5.3 Typical continuity test box.

the earth pin of socket outlets and the female part of plugs. Poking a small screwdriver into the earth socket is not good working practice.

For Class I cord sets, why not use the arrangement shown in Figure 5.3 and add a selection of recessed sockets to house the range of female plugs found on cord sets? All their earth pins would be connected to the earth stud. For extension leads incorporating a socket or sockets, use the earth pin from an old plug, as this is designed to enter the earth pin socket.

Insulation resistance

Realistically, this test can only be carried out on Class I equipment. It is made to ensure that there is no breakdown of insulation between the protective earth and live (line and neutral) parts of the appliance and its lead.

For Class II items, there are no earthed parts and one test probe would need to be placed at various points on the body of the appliance in order to check the integrity of the casing.

Items that have a cord set (e.g. a kettle) should have the cord set plugged into the appliance and the appliance switch should be in the 'on' position.

There are two tests that can be made, using either the applied voltage method or the earth leakage method.

The applied voltage method

This is conducted using an insulation resistance tester, set on 500 V DC. The test is made between the line and neutral *connected together*, and the protective earth. (For three-phase items, all live conductors are connected together.) This is best achieved using the same arrangement as shown in Figure 5.2, but with the addition of a line/neutral stud connected to the socket's line and neutral (Figure 5.4).

Care must be taken when conducting this test to ensure that the appliance is not touched during the process. Also, it should be noted that some items of equipment have filter networks connected across line and earth

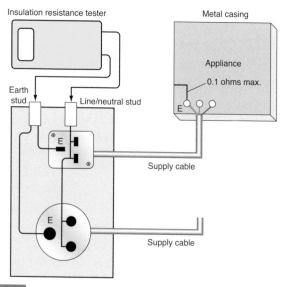

FIGURE 5.4 Typical continuity and insulation test box.

Table 5.2   Insulation Resistance Values

Appliance Class	Insulation Resistance
Class I heating equipment ≥ 3 kW	0.3 MΩ
General Class I equipment	1 MΩ
Class II equipment	2 MΩ
Class III equipment	250 kΩ

terminals and this may lead to unduly low values. The values recorded should not be less than those shown in Table 5.2.

The earth leakage method

This is achieved using a portable appliance tester that subjects the insulation to a less onerous voltage (usually 250 V) than that delivered by an insulation resistance tester. Here, the leakage current across the insulation is measured, and appliance testers usually set the maximum value at 3.5 mA.

Whichever method is used, there is a chance of pessimistically low values occurring when some heating or cooking appliances are tested. This is usually due to moisture seeping into the insulation of the elements. In this case, it is wise to switch such equipment on for a short while to dry the elements out before testing.

> **Note**
>
> Many portable appliance testers have the facility to conduct a 'dielectric strength' or 'flash' test, which is basically an insulation resistance test at 1250 V for Class I equipment and 3570 V for Class II. Such voltages could cause damage to insulation and should *not* be carried out for in-service tests.

Touch current

This is an alternative to insulation resistance testing and is only available on the more expensive/comprehensive types of PAT tester. The exact method of conducting this test is not at all clear in The Code of Practice for In-Service Inspection and Testing of Electrical Equipment, and as it is quite unusual to perform such a test in normal circumstances it has been omitted from this volume.

Functional checks

If testing has been carried out using separate instruments, just switch the equipment to ensure that it is working. If a portable appliance tester is used, there is usually a facility for conducting a 'load test'. The equipment is automatically switched on and the power consumption is measured while the item is on load. This is useful as it indicates if the equipment is working to its full capacity, for example a 2 kW reading on a 3 kW heater suggests a broken element.

Questions

1. Why should inspection of equipment always precede testing?
2. Which instruments are recognized for carrying out tests on electrical equipment?
3. For Class I equipment, what is the maximum resistance permissible between the equipment earth terminal and the casing?
4. What is the advantage of testing the earth continuity of an item via its supply lead and what action should be taken regarding the resistance of this lead?
5. Extension leads should be tested as what Class of equipment?
6. What test is performed to establish the condition of the insulation between live and earth conductors?
7. State the methods that can be used to perform the test in question 6 above.
8. Having completed the instrument tests required on equipment, what further test should be conducted?

Answers

1. Inspection may reveal serious faults that testing will not pick up.
2. Portable Appliance Tester or individual Continuity and Insulation Resistance testers.
3. 0.1 Ω.
4. It checks the continuity of the cpc in the lead. Lead resistance should be deducted from the test result.
5. Class I.
6. Insulation Resistance test.
7. Applied voltage and earth leakage tests.
8. Functional or load test.

Shock Risk

Important terms used in this appendix:

- Basic protection
- Fault protection

By the end of this appendix the reader should:

- have an understanding of the effects of electric current on the body,
- know how shock risk is reduced,
- be aware of the importance of earthing.

As we have seen in Chapter 1, all who are involved with electrical systems are 'Duty holders' in Law. For those operatives who have only a limited knowledge of electricity, but are nevertheless charged with their company's appliance testing, an understanding of electric shock will help to give more meaning and confidence to the inspection and test process.

ELECTRIC SHOCK

This is the passage of current through the body of such magnitude as to have significant harmful effects. Table A1.1 and Figure A1.1 illustrate the generally accepted effects of current passing through the human body. How, then, are we at risk of electric shock and how do we protect against it?

Table A1.1 Effects of current passing through the human body

1–2 mA	Barely perceptible, no harmful effects
5–10 mA	Throw off, painful sensation
10–15 mA	Muscular contraction, can't let go
20–30 mA	Impaired breathing
50 mA and above	Ventricular fibrillation and death

PAT: Portable Appliance Testing: In-Service Inspection and Testing of Electrical Equipment. 978-1-138-84929-7. © Brian Scaddan. Published by Taylor & Francis. All rights reserved.

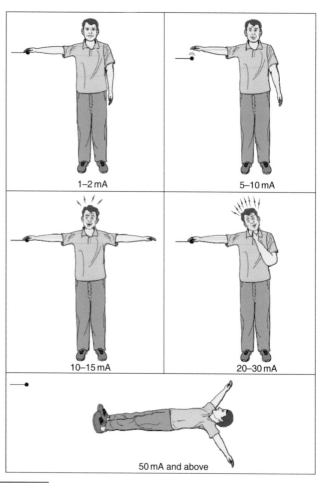

1–2 mA

5–10 mA

10–15 mA

20–30 mA

50 mA and above

FIGURE A1.1   Electric shock levels.

These are two ways in which we can be at risk:

1. Touching live parts of equipment or systems that are intended to be live.
2. Touching conductive parts which are not meant to be live, but which have become live due to a fault.

The conductive parts associated with point 2 above can be either metal-work of electrical equipment and accessories (Class I) and that of electrical wiring systems (e.g. metal conduit and trunking), called *exposed conductive* parts, or other metalwork (e.g. pipes, radiators and girders), called *extraneous conductive* parts.

BASIC PROTECTION

How can we prevent danger to persons and livestock from contact with intentionally live parts? Clearly we must minimize the risk of such contact and this can be achieved by *basic* protection, which comprises:

- insulating any live parts
- ensuring any uninsulated live parts are housed in suitable enclosures and/or are behind barriers.

The use of a residual current device (RCD) cannot prevent such contact, but it can be used as additional protection to any of the other measures taken, provided that it is rated at 30 mA or less and has a tripping time of not more than 40 ms at an operating current of 150 mA.

It should be noted that RCDs are not the panacea for all electrical ills, they can malfunction, but they are a valid and effective backup to the other methods. They must not be used as the sole means of protection.

FAULT PROTECTION

How, under single fault conditions, can we protect against shock from contact with live, exposed or extraneous conductive parts whilst touching earth, or from contact between live exposed and/or extraneous conductive parts? The most common method is by protective earthing and protective equipotential bonding and automatic disconnection of supply.

All extraneous conductive parts are joined together with a main protective bonding conductor and connected to the main earthing terminal, and all exposed conductive parts are connected to the main earthing terminal by the circuit protective conductors. Add to this, overcurrent protection that will operate fast enough when a fault occurs and the risk of severe electric shock is significantly reduced.

WHAT IS EARTH AND WHY AND HOW DO WE CONNECT TO IT?

The thin layer of material which covers our planet – rock, clay, chalk or whatever – is what we in the world of electricity refer to as earth. So, why do we need to connect anything to it? After all, it is not as if earth is a good conductor.

It might be wise at this stage to investigate potential difference (PD). A PD is exactly what it says it is: a difference in potential (volts). In this way, two conductors having PDs of, say, 20 and 26 V have a PD between them of $26 - 20 = 6$ V. The original PDs (i.e. 20 and 26 V) are the PDs between 20 V and 0 V and 26 V and 0 V. So where does this 0 V or zero potential come from? The simple answer is, in our case, the earth. The definition of earth is, therefore, the conductive mass of earth, whose electric potential at any point is conventionally taken as zero.

Thus, if we connect a voltmeter between a live part (e.g. the line conductor of a socket outlet) and earth, we may read 230 V; the conductor is at 230 V and the earth at zero. The earth provides a path to complete the circuit. We would measure nothing at all if we connected our voltmeter between, say, the positive 12 V terminal of a car battery and earth, as in this case the earth plays no part in any circuit.

Figure A1.2 illustrates this difference.

So, a person in an installation touching a live part whilst standing on the earth would take the place of the voltmeter and could suffer a severe electric shock. Remember that the accepted lethal level of shock current passing through a person is only 50 mA or 1/20 A. The same situation would arise if the person were touching a faulty appliance and a gas or water pipe (Figure A1.3).

One method of providing some measure of protection against these effects is, as we have seen, to join together (bond) all metallic parts and connect them to earth. This ensures that all metalwork in a healthy installation is at or near 0 V and, under fault conditions, all metalwork will rise to a similar potential. So, simultaneous contact with two such metal parts would not result in a dangerous shock, as there would be no significant PD between them.

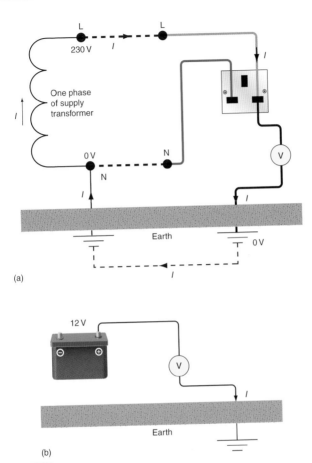

(a)

(b)

FIGURE A1.2   (a) Return path via earth; (b) No return path via earth.

Unfortunately, as mentioned, earth itself is not a good conductor, unless it is very wet. Therefore, it presents a high resistance to the flow of fault current. This resistance is usually enough to restrict fault current to a level well below that of the rating of the protective device, leaving a faulty circuit uninterrupted. Clearly this is an unhealthy situation, and an RCD would be needed for earth fault protection.

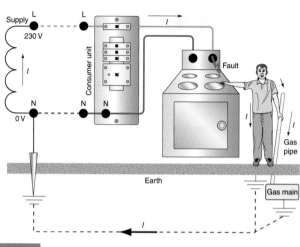

FIGURE A1.3 Electric shock path.

In all but the most rural areas, consumers can connect to a metallic earth return conductor, which is ultimately connected to the earthed neutral of the supply. This, of course, presents a low-resistance path for fault currents to operate the protection.

In summary, connecting metalwork to earth places that metal at or near zero potential and bonding between metallic parts puts such parts at a similar potential even under fault conditions. Add to this a low-resistance earth fault return path, which will enable the circuit protection to operate very fast, and we have significantly reduced the risk of electric shock. We can see from this how important it is to check that equipment earthing is satisfactory and that there is no damage to conductor insulation.

Questions

1. What is considered to be the lethal level of electric shock?
2. What two means of protection may be taken to reduce the risk of electric shock?
3. What is the potential of the mass of earth?
4. Why is an unearthed item of Class I equipment dangerous?

Answers

1. 50 mA.
2. Basic and fault.
3. Zero volts.
4. There is no return path for earth fault currents to operate protective devices and faulty equipment will stay live giving rise to shock risk.

Basic Electrical Theory Revision

Important terms used in this appendix:

- Current
- Voltage
- Resistance
- Power

By the end of this appendix the reader should:

- be able to carry out a simple calculation involving current, voltage, resistance and power,
- be able to determine conductor resistance given their sizes.

This appendix has been added in order to jog the memory of those who have some electrical background and to offer a basic explanation of theory topics within this book for those relatively new to the subject.

ELECTRICAL QUANTITIES AND UNITS

Quantity	Symbol	Units
Current	I	Ampere (A)
Voltage	V	Volt (V)
Resistance	R	Ohm (Ω)
Power	P	Watt (W)

Current

This is the flow of electrons in a conductor.

Voltage

This is the electrical pressure causing the current to flow.

PAT: Portable Appliance Testing: In-Service Inspection and Testing of Electrical Equipment.
978-1-138-84929-7. © Brian Scaddan. Published by Taylor & Francis. All rights reserved.

Resistance

This is the opposition to the flow of current in a conductor determined by its length, cross-sectional area and temperature.

Power

This is the product of current and voltage, hence $P = I \times V$.

RELATIONSHIP BETWEEN VOLTAGE, CURRENT AND RESISTANCE

- Voltage = Current × Resistance; $V = I \times R$,
- Current = Voltage/Resistance; $I = V/R$, or
- Resistance = Voltage/Current; $R = V/I$.

COMMON MULTIPLES OF UNITS

Current, I (amperes)	kA	mA
	Kilo-amperes	Milli-amperes
	1000 amperes	1/1000 of an ampere
Voltage, V (volts)	kV	mV
	Kilovolts	Millivolts
	1000 volts	1/1000 of a volt
Resistance, R (ohms)	MΩ	mΩ
	Megohms	Milli-ohms
	1 000 000 ohms	1/1000 of an ohm
Power, P (watts)	MW	kW
	Megawatt	Kilowatt
	1 000 000 watts	1000 watts

RESISTANCE IN SERIES

These are resistances joined end to end in the form of a chain. The total resistance increases as more resistances are added (Figure A2.1).

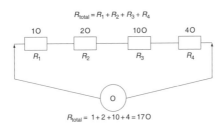

$$R_{total} = R_1 + R_2 + R_3 + R_4$$

$$R_{total} = 1 + 2 + 10 + 4 = 17\,\Omega$$

FIGURE A2.1 Resistances in series.

Hence, if a cable length is increased, its resistance will increase in proportion. For example, a 100 m length of conductor has twice the resistance of a 50 m length of the same conductor.

RESISTANCE IN PARALLEL

These are resistances joined like the rungs of a ladder. Here the total resistance decreases with a greater number of rungs (Figure A2.2).

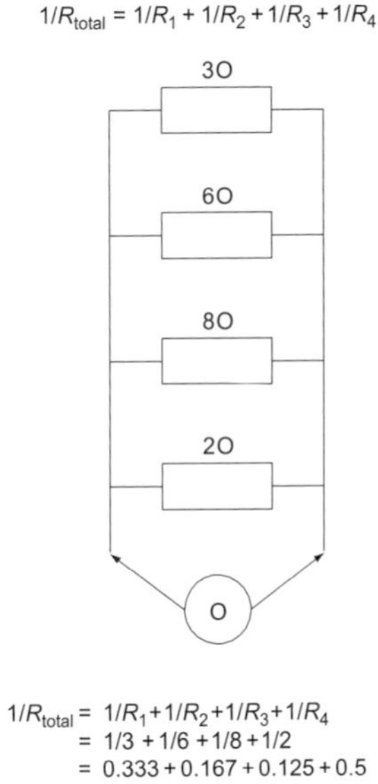

$$1/R_{total} = 1/R_1 + 1/R_2 + 1/R_3 + 1/R_4$$

$$
\begin{aligned}
1/R_{total} &= 1/R_1 + 1/R_2 + 1/R_3 + 1/R_4 \\
&= 1/3 + 1/6 + 1/8 + 1/2 \\
&= 0.333 + 0.167 + 0.125 + 0.5
\end{aligned}
$$

FIGURE A2.2 Resistances in parallel.

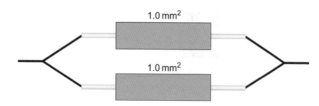

FIGURE A2.3 Conductors in parallel.

The insulation between conductors is in fact countless millions of very high value resistances in parallel. Hence, an increase in cable length results in a decrease in insulation resistance. This value is measured in millions of ohms, that is megohms (MΩ).

The overall resistance of two or more conductors will also decrease if they are connected in parallel (Figure A2.3).

The total resistance will be half of either one and would be the same as the resistance of a 2 mm^2 conductor. Hence, resistance decreases if conductor cross-sectional area increases.

Example A2.1

If the resistance of a 1.0 mm^2 conductor is 19.5 mΩ/m, what would be the resistance of

1. 5 m of 1.0 mm^2 conductor?
2. 1 m of 6.0 mm^2 conductor?
3. 25 m of 4.0 mm^2 conductor?
4. 12 m of 0.75 mm^2 conductor?

Answers

1. 5 m × 19.5 mΩ/m = 0.0975 Ω.
2. A 6.0 mm^2 conductor would have a resistance 6 times less than a 1.0 mm^2 conductor (i.e. 19.5/6 = 3.25 mΩ).
3. 25 m of 4.0 mm^2 would be 19.5 × 25/4 × 1000 = 0.12 Ω.
4. 19.5 mΩ/m × 1.5 (the ratio of 0.75 mm^2 to 1.00 mm^2 conductor) × 12 m = 0.351 Ω.

Questions

1. Calculate the resistance of a 230V appliance which draws 10A from the supply.
2. Calculate the current drawn by a 230V appliance that has a resistance of 11.5Ω.
3. What is the kW rating of a 230V, 20A appliance?

Answers

1. 23Ω.
2. 20A.
3. 4.6kW.

Sample 2377 Questions

THE MANAGEMENT OF ELECTRICAL EQUIPMENT MAINTENANCE

1. Which one of the following is a statutory document?
   (a) A British Standard
   (b) IET Wiring Regulations
   (c) IET Codes of Practice
   (d) Electricity at Work Regulations.

2. Which one of the following regulations states: 'Every employer shall make a suitable and sufficient assessment of the risk to the health and safety of his or her employees and to persons not in his or her employment'?
   (a) The Electricity Safety, Quality and Continuity Regulations 2002
   (b) The Electricity at Work Regulations
   (c) The Provision and Use of Work Equipment Regulations
   (d) The Management of Health and Safety at Work Regulations.

3. Certain sections of the Health and Safety at Work Regulations put a duty of care upon:
   (a) employees only
   (b) employers only
   (c) both employees and the general public
   (d) both employers and employees.

4. Which one of the following regulations state: 'As may be necessary to prevent danger, all systems shall be maintained so as to prevent, so far as is reasonably practicable, such danger'?
   (a) The Electricity at Work Regulations
   (b) The IET Wiring Regulations
   (c) The Provision and Use of Work Equipment Regulations
   (d) The Management of Health and Safety at Work Regulations.

5. The scope of legislation of inspection and testing of electrical equipment extends to distribution systems up to:
   (a) 230 V
   (b) 400 V
   (c) 11 kV
   (d) 400 kV.

6. The Code of Practice for In-service Inspection and Testing of Electrical Equipment does **not** apply to:
   (a) shops
   (b) offices
   (c) caravan sites
   (d) petrol station forecourts.

7. The safety and proper functioning of certain portable appliances and equipment depends on the integrity of the fixed installation. Requirements for the inspecting and testing of fixed installations are given in:
   (a) BS 2754
   (b) BS 7671
   (c) BS EN 60947
   (d) BS EN 60898.

8. Transportable equipment is sometimes called:
   (a) hand-held appliance or equipment
   (b) stationary equipment or appliance
   (c) moveable equipment
   (d) portable appliance.

9. An electric toaster is classified as:
   (a) a portable appliance
   (b) moveable equipment
   (c) a hand-held appliance
   (d) equipment for 'building in'.

10. Which one of the following domestic electrical appliances may be regarded as an item of stationary equipment?
    (a) A bathroom heater
    (b) A visual display unit

(c) A washing machine

(d) A built-in electric cooker.

11. A portable appliance that is supplied by a flexible cable incorporating a protective conductor is classified as:

(a) Class I

(b) double insulated

(c) metal clad Class II

(d) Class III.

12. Stationary equipment/appliances are defined as not being provided with a carrying handle and have a mass exceeding:

(a) 10 kg

(b) 12 kg

(c) 15 kg

(d) 18 kg.

13. A freezer is classified as:

(a) a stationary appliance or equipment

(b) a hand-held appliance or equipment

(c) moveable equipment

(d) a portable appliance.

14. A BS 3535 safety isolating transformer having a voltage not exceeding 50 V is used to supply certain equipment. The class of such equipment is:

(a) Class 0

(b) Class I

(c) Class II

(d) Class III.

15. Which size of the following three-core extension leads is too large for a standard 13 A plug?

(a) 2.5 mm^2

(b) 1.5 mm^2

(c) 1.25 mm^2

(d) 1.00 mm^2.

16. Which one of the following arrangements would **not** meet the requirements of the IET Code of Practice?
    (a) Class I equipment supplied by a 1.5 mm² three-core extension lead connected into a 13 A three-pin socket outlet.
    (b) Class II equipment supplied by a 1.5 mm² two-core extension lead connected into a 13 A three-pin socket outlet.
    (c) Class I equipment supplied by a 2.5 mm² three-core extension lead connected into a BS EN 60309-2 socket outlet.
    (d) Class III equipment supplied by a two-core flexible cable connected into the secondary of an isolating transformer supplying SELV lighting equipment.

17. Which one of the following size and length extension leads should be used in conjunction with an RCD used for supplementary protection?
    (a) 1.5 mm², 10 m long
    (b) 1.5 mm², 15 m long
    (c) 2.5 mm², 20 m long
    (d) 2.5 mm², 30 m long.

18. During the inspection and testing process, which of the following is not required?
    (a) Preliminary inspection
    (b) Earth continuity tests (for Class I equipment)
    (c) Insulation testing
    (d) Earth continuity test on Class II equipment.

19. Which one of the following would **not** be conducted during routine inspection and testing of appliances?
    (a) Preliminary inspection
    (b) Earth continuity tests
    (c) Type testing
    (d) Functional checks.

20. When performing in-service testing on Class I equipment, which one of the following is **not** required?
    (a) Type testing to a British Standard
    (b) Earth continuity test
    (c) Insulation testing
    (d) Functional checks.

21. Details of which of the following must be recorded when carrying out a safety check on an electrical appliance?
    (a) Manufacturer's name and address
    (b) Combined inspection and test
    (c) User check revealing no damage to equipment
    (d) Applicable British Standards.

22. Which one of the following will **not** affect the frequency of inspection and testing for an electrical appliance?
    (a) The integrity of the fixed electrical installation
    (b) Environment in which it is to be used
    (c) The user
    (d) The equipment class.

23. Recorded testing but not inspecting of equipment may be omitted if the:
    (a) equipment is of Class I construction and in a low-risk area
    (b) equipment is of Class II construction and in a low-risk area
    (c) user of the equipment reports damage as and when it becomes evident
    (d) equipment is a hand-held appliance.

24. The table of suggested frequency of inspection and testing for electrical equipment gives details of:
    (a) the forms required for such testing
    (b) maximum and minimum values of test results
    (c) the required sequence of visual checks to be made
    (d) types of premises within which electrical equipment is operated and user check requirements.

25. The suggested initial frequency for a formal visual inspection of a hand-held Class II electric iron in a hotel is:
    (a) 1 month
    (b) 6 months
    (c) 12 months
    (d) 24 months.

26. The suggested frequency for user checks for children's rides in a fairground is:
    (a) weekly
    (b) monthly
    (c) daily
    (d) 12 months.

27. Which one of the following tests should **not** be applied routinely to equipment?
    (a) Earth continuity
    (b) Insulation resistance
    (c) Polarity
    (d) Dielectric strength.

28. The first electrical test to be applied to Class I equipment is:
    (a) insulation resistance
    (b) earth continuity
    (c) dielectric strength
    (d) polarity.

29. When information regarding test procedures is unavailable from the manufacturer or supplier of IT equipment, which one of the following electrical tests should **not** be undertaken?
    (a) Earth continuity
    (b) Polarity
    (c) Functional
    (d) Insulation.

30. The purpose of an equipment register is to ensure:
    (a) compliance with the Electricity at Work Regulations
    (b) that maintenance procedures are recorded
    (c) the frequency of inspection and test is reviewed
    (d) inspection and testing is performed.

31. Identification of all electrical equipments within a duty holder's control is required in order to produce:
    (a) 'pass' safety check equipment label
    (b) faulty equipment register
    (c) equipment register
    (d) repair register.

32. Which one of the following items of information is **not** required on an inspection and test label?
    (a) An indication of whether the equipment has passed or failed the safety tests
    (b) Details of previous test results
    (c) Date at time of testing
    (d) Appliance or equipment number.

33. All electrical equipments should be marked with a unique serial number to aid:
    (a) disconnection
    (b) identification
    (c) risk assessment
    (d) interpretation of test results.

34. Information provided for equipment which requires routine inspection and/or testing should consist of:
    (a) an indelible bar-code system
    (b) an identification code to enable the equipment to be uniquely identifiable
    (c) operating instructions regarding the test equipment
    (d) an indication of the results which may be expected during inspections and/or tests.

35. Which one of the following is **not** required to be tested within the scope of the IET Code of Practice?
    (a) Fixed equipment
    (b) Fixed installations
    (c) Electrical tools
    (d) Portable appliances.

36. The Memorandum of Guidance on the Electricity at Work Regulations 1989 advises that equipment records:
    (a) should be kept throughout the working life of the equipment
    (b) should only be kept where the equipment is used in high-risk areas
    (c) are not required where the equipment is used in low-risk areas
    (d) are not required if the equipment is fed from a 110 V safety supply.

37. Records of all maintenance activities relating to electrical appliances must be kept, including details of the:
    (a) initial cost
    (b) procurement of equipment
    (c) estimated replacement date
    (d) estimated replacement cost.

38. The person responsible for carrying out an inspection and test on an appliance should have made available to them:
    (a) a list of all the users of equipment
    (b) a copy of the Electricity at Work Regulations
    (c) a copy of the Health and Safety at Work Act
    (d) previous inspection and test results.

39. Which voltage must be used when carrying out an insulation resistance test on a Class I toaster?
    (a) 3750 V a.c.
    (b) 500 V d.c.
    (c) 1000 V d.c.
    (d) 500 V a.c.

40. An insulation resistance tester should be capable of:
    (a) delivering a minimum voltage of 1000 V d.c. to the load
    (b) testing the continuity of an earthing circuit
    (c) delivering a maximum voltage of 25 A through the load
    (d) maintaining the test voltage required across the load.

41. Where a user check reveals damage to the equipment, it must be reported to:
    (a) the equipment manufacturer
    (b) the Health and Safety Inspectorate
    (c) a responsible person
    (d) a manager of an inspection and test organization.

42. The manager of an inspection and test organization should be able to:
    (a) repair faulty electrical equipment
    (b) instruct untrained persons in the use of portable appliance testers

(c) know their legal responsibilities under the Electricity at Work Regulations

(d) demonstrate competence in the use of appliance testers.

43. Which one of the following is outside the scope of the IET Code of Practice for Inspection and Testing of In-Service Electrical Equipment?
    (a) Those who inspect and test
    (b) The user of electrical appliances
    (c) Managers of the inspection and test organization
    (d) The hirer of electrical portable appliances and equipment.

44. Earth continuity testing may in certain circumstances be carried out by means of:
    (a) a low-resistance ohmmeter
    (b) an insulation resistance tester
    (c) a bell set and battery
    (d) an instrument complying with BS EN 60309.

45. Test leads and probes used to measure voltages over 50V a.c. and 100V d.c. should comply with:
    (a) BS 7671
    (b) Health and Safety Executive Guidance Note GS 38
    (c) BS 5490 Specification for Classification of Protection
    (d) IEC Publication 479.

INSPECTION AND TESTING OF ELECTRICAL EQUIPMENT

1. Which one of the following sections of the Health and Safety at Work Act places a duty of care on employees?
   (a) 2
   (b) 3
   (c) 6
   (d) 7.

2. The scope of the IET Code of Practice for In-service Inspection and Testing of Electrical Equipment includes d.c. voltages between conductors and earth up to:
   (a) 1500V d.c.
   (b) 1000V d.c.

   (c) 900V d.c.

   (d) 600V d.c.

3. Which one of the following premises is outside the scope of the Health and Safety at Work Act?
   (a) Leased offices or shops
   (b) Privately owned stores
   (c) Owner occupied dwellings
   (d) Light industrial units.

4. In which one of the following documents can HSE guidance on the maintenance of portable and transportable electrical equipment be found?
   (a) HSG 107
   (b) HSG 85
   (c) HSG 141
   (d) HSG 118.

5. Which one of the following is the main reason for in-service inspection and testing of electrical equipment?
   (a) To ascertain whether maintenance is required
   (b) Carrying out stock control
   (c) Monitoring the hiring of equipment
   (d) Ensuring equipment is the current model.

6. Which one of the following items is not within the scope of the IET Code of Practice for In-service Inspection and Testing of Electrical Equipment?
   (a) Hand held electrical appliances
   (b) Moveable electrical appliances
   (c) Office IT equipment
   (d) Circuit breakers in consumer units.

7. Which one of the following symbols represents resistance?
   (a) Z
   (b) Ω
   (c) A
   (d) μ.

8. Which one of the following is the unit of measurement for frequency?
   (a) Amperes
   (b) Hertz
   (c) Ohms
   (d) Volts.

9. Which one of the following values is another way of representing 30 mA?
   (a) 30 A
   (b) 0.3 A
   (c) 0.03 A
   (d) 0.003 A.

10. Which one of the following equipment types is a tumble dryer on wheels and supplied by a 13 A plug?
    (a) Moveable
    (b) Portable
    (c) Stationary
    (d) Hand-held

11. Which one of the following devices should be installed for portable equipment used outdoors?
    (a) An overload relay
    (b) An inverter
    (c) A residual current device
    (d) A surge protection device.

12. Which one of the following classification symbols is used to denote Class II equipment?
    (a) ☐
    (b) ◯
    (c) II
    (d) ◇

13. Which one of the following situations may require protection against electric shock by fault protection?
    (a) A broken cover on a 13 A plug
    (b) A badly twisted appliance cable
    (c) A line to earth fault on Class I equipment
    (d) A line to earth fault on Class II equipment.

14. Which one of the following indicates how protection against electric shock under fault free conditions is achieved?
    (a) Earth protection
    (b) Basic protection
    (c) Supplementary protection
    (d) Fault protection.

15. Which one of the following would be the effect on a conductor's resistance if its cross-sectional-area were doubled?
    (a) No change
    (b) Halved
    (c) Quadrupled
    (d) Doubled.

16. Which one of the following would be the effect on a conductor's resistance if its length were doubled?
    (a) No change
    (b) Halved
    (c) Quadrupled
    (d) Doubled.

17. An appliance lead has a measured value of earth continuity resistance of $0.4\,\Omega$. Which one of the following would be the value of resistance if the lead length were doubled?
    (a) $0.2\,\Omega$
    (b) $0.4\,\Omega$
    (c) $0.8\,\Omega$
    (d) $1.6\,\Omega$.

18. A $1.25\,mm^2$ appliance lead is $2.5\,m$ long. Which one of the following would be the resistance of the cpc at a temperature of 20°C?
    (a) $39.0\,\Omega$
    (b) $31.2\,\Omega$

(c) 0.039 Ω
(d) 0.031 Ω.

19. Which one of the following has a direct effect on the resistance of the cpc in an appliance cable?
   (a) The total number of conductors in the appliance cable
   (b) The cross-sectional-area of the cpc
   (c) The outside diameter of the appliance cable
   (d) The thickness of the cpc insulation material.

20. Under short circuit fault conditions, an appliance fuse must disconnect quickly. Which one of the following would cause longer disconnection times for a Class II appliance?
   (a) The appliance lead cross-sectional-area is increased
   (b) The appliance lead length is increased
   (c) The appliance lead length is decreased
   (d) The appliance lead cpc cross-sectional-area is decreased.

21. Which one of the following actions should be taken where it is unlikely that users of new equipment will report any abuse of the equipment?
   (a) Extend the frequency between inspections and testing
   (b) Take no action as the user is responsible
   (c) Reduce the frequency between inspections and testing
   (d) Ensure the user records all user checks.

22. In which one of the following environments is there no requirement to carry out and record combined inspection and testing of Class II equipment?
   (a) Schools
   (b) Construction sites
   (c) Commercial kitchens
   (d) Offices and shops.

23. Which one of the following actions is applicable to user checks on electrical equipment?
   (a) Record all user checks and inform responsible person
   (b) Only to be conducted by electrically competent persons
   (c) Only record user checks when faults are found
   (d) Record all user checks on the repair register.

24. Which one of the following set of actions is required for formal visual inspections?
    (a) Inspection without tests, satisfactory and unsatisfactory results recorded
    (b) Inspection with tests, satisfactory and unsatisfactory results recorded
    (c) Inspection without tests, only unsatisfactory results recorded
    (d) Inspection without tests, only satisfactory results recorded.

25. Which one of the following pairs of factors would influence the decision on the frequency of inspection and testing of equipment?
    (a) Equipment current rating and equipment type
    (b) Equipment type and the environment
    (c) Equipment fuse rating and environment
    (d) Equipment current rating and construction.

26. Which one of the following is the recommended initial frequency for formal visual inspection of Class I portable equipment in a hotel bedroom?
    (a) 24 months
    (b) 12 months
    (c) 6 months
    (d) weekly.

27. Which one of the following actions should be taken where recurring damage to equipment is noted?
    (a) Keep replacing the equipment with a similar type
    (b) Decrease the frequency of inspection and testing
    (c) Increase the frequency of inspection and testing
    (d) Return the equipment to use after each repair.

28. Which one of the following is the most important check that can be made on an item of equipment?
    (a) Instrument tests
    (b) User checks
    (c) Manufactured to specification
    (d) Visual inspection.

29. Which one of the following is the recommended percentage value of instrument accuracy above which an instrument should be recalibrated?
    (a) 2%
    (b) 5%
    (c) 10%
    (d) 15%.

30. It is recommended that test instrument accuracy is maintained. This may be achieved by:
    (a) more frequent use of the instrument
    (b) less frequent use of the instrument
    (c) renewing the instrument warranty
    (d) maintaining instrument records.

31. Which one of the following is the most economic way of ensuring the accuracy of test instruments?
    (a) Return the instrument to the manufacturer for calibration every quarter
    (b) Check the instrument frequently against known values and record results
    (c) Check the instrument on various appliances that have known faults
    (d) Change instrument batteries before and after use.

32. Which one of the following tests **cannot** be performed when testing an item of Class II equipment?
    (a) Earth continuity
    (b) Insulation resistance
    (c) Polarity
    (d) Functional.

33. Which one of the following is not checked whilst inspecting a Class I electric toaster?
    (a) Condition of casing
    (b) Condition of supply lead
    (c) Earth continuity
    (d) Rating of plug fuse.

34. A 3 kW convector heater has three elements. Which one of the following readings obtained during a functional test would indicate that one element was open circuit?
    (a) 9000 W
    (b) 3000 W
    (c) 2000 W
    (d) 1000 W.

35. Which one of the following is the reason for a polarity check on equipment?
    (a) To ensure earth continuity
    (b) To ensure L & N connections are correct
    (c) To check the condition of the supply lead
    (d) To ensure the equipment functions correctly.

36. Which one of the following items would require a polarity test to be carried out?
    (a) Multiway adaptor
    (b) Extension lead
    (c) Plug fuse
    (d) Fluorescent tube.

37. An insulation resistance test is carried out in order to:
    (a) establish the resistance of the cpc of the equipment lead
    (b) determine the Class of equipment when it is not known
    (c) establish that the equipment insulation is to specification
    (d) check that there is no damage/deterioration in the insulation.

38. During an insulation resistance test the L and N conductors should be:
    (a) connected together
    (b) separated
    (c) bonded to earth
    (d) tested between them.

39. Which one of the following is the minimum value of insulation resistance for an item of Class I heating and cooking equipment with a rating ≥ 3 kW?
    (a) 25 kΩ
    (b) 2.0 MΩ

(c) 1.0 MΩ

(d) 0.3 MΩ.

40. Where IT equipment may be vulnerable to the standard insulation resistance test this test may be replaced by:

    (a) a 'hard test'

    (b) an 'alternative test'

    (c) a 'soft test'

    (d) a 'medium test'.

41. The protective conductor 'touch current' is to be measured for portable or hand-held Class I equipment. The maximum value should not exceed:

    (a) 5 mA

    (b) 3.5 mA

    (c) 0.75 mA

    (d) 0.25 mA.

42. When tested at 100% of its rated residual operating current, an RCD to BS EN 61008 should disconnect in a maximum time of:

    (a) 30 ms

    (b) 200 ms

    (c) 300 ms

    (d) 500 ms.

43. Where an extension lead is fitted with an RCD, the RCD must have a maximum residual operating current of:

    (a) 30 mA

    (b) 20 mA

    (c) 10 mA

    (d) 5 mA.

44. Which one of the following is checked by operating the test button on an RCD?

    (a) Correct polarity of the RCD

    (b) The correct operation

    (c) The disconnection time is correct

    (d) Its electrical efficiency.

45. When testing an appliance with a detachable 3-core lead set, the lead set should be tested:
(a) separately as a Class II appliance
(b) with the appliance connected
(c) separately as a Class I appliance
(d) separately as a Class III appliance.

46. Which one of the following is the maximum length of a $1.5\,mm^2$ extension lead?
(a) 5 m
(b) 12 m
(c) 15 m
(d) 25 m.

47. Which one of the following should be referred to when equipment leakage currents exceed 10 mA?
(a) GS 38
(b) GN3
(c) EAWR
(d) BS7671.

48. Which one of the following items of documentation would not be maintained by an organisation carrying out in-service inspection and testing?
(a) Faulty equipment register
(b) User register
(c) Repair register
(d) Equipment register.

49. Which one of the following forms should be used after damaged equipment has been removed from service?
(a) V .5
(b) V .4
(c) V .3
(d) V .2.

50. An equipment register, form VI.1, must state the name of the
(a) competent person
(b) inspector and tester

(c) safety officer

(d) responsible person.

ANSWERS

The management of electrical equipment maintenance

1. d	2. d	3. d	4. a	5. d
6. d	7. b	8. c	9. a	10. c
11. a	12. d	13. a	14. d	15. a
16. b	17. d	18. d	19. c	20. a
21. c	22. a	23. b	24. d	25. b
26. c	27. d	28. b	29. d	30. c
31. c	32. b	33. b	34. b	35. b
36. a	37. b	38. d	39. b	40. d
41. c	42. c	43. d	44. a	45. b

Inspection and testing of electrical equipment

In brackets you will see where to find the answers in the Code of Practice.

1. d   (3.1.1 on Page 33)
2. c   (1.4 on Page 20)
3. c   (1.3 on Page 20)
4. a   (App IV on Page 122)
5. a   (1.5 on Page 21)
6. d   (1.2 on Page 20)
7. b
8. b
9. c
10. a   (Page 30 or Page 43)
11. c
12. a   (Fig. 11.12 on Page 78)
13. c   (Page 27)
14. b   (Page 23)
15. b   (App VI on Page 134)

16. d   (App VI on Page 134)
17. c   (App VI on Page 134)
18. c   (App VI on Page 134)
19. b   (App VI on Page 134)
20. b
21. c   (7.3 on Page 50)
22. d   (Table 7.1 on Page 52)
23. c   (Page 80)
24. a   (6.4 on Page 48)
25. b   (Table 7.1 on Page 52)
26. b   (Table 7.1 on Page 52)
27. c   (Page 51)
28. d   (Page 49)
29. b   (10.5 on Page 67)
30. d   (10.5 on Page 67)
31. b   (10.5 on Page 67)
32. a   (15.4 on Page 91)
33. d   (7.2 on Page 50)
34. c
35. b   (15.10 on Page 98)
36. b   (15.10 on Page 97–8)
37. d
38. a   (15.5 on Page 93)
39. d   (Table 15.2 on Page 94)
40. c   (15.6 on Page 95)
41. c   (Table 15.3 on Page 96)
42. c   (Table 15.5 on Page 99)
43. a   (15.10.2 on Page 98)
44. b   (15.10.2 on Page 98)
45. c   (15.9 on Page 97)
46. c   (Table 15.4 on Page 98)
47. d   (15.11 on Page 100)
48. b   (App V on Page 125)
49. a   (App V on Page 125)
50. d   (App V on Page 125)

Index